This is an island in the ocean. Let's look and listen!

The waves hit the sand in slow motion.

Do the plants look strange to you?

Whose tail is that? This animal likes action!

This section of the island has birds.

This section has little shops.

One shop was once a ship!

What should you do on this island?
Use your eyes and ears!